YOUR KNOWLEDGE HAS VALUE

- We will publish your bachelor's and
  master's thesis, essays and papers

- Your own eBook and book -
  sold worldwide in all relevant shops

- Earn money with each sale

Upload your text at www.GRIN.com
and publish for free

Akhilesh Kumar

Jatropha curcas: A potential genetic resource for herbal medicine and liquid bio-fuel

GRIN Publishing

Bibliographic information published by the German National Library:

The German National Library lists this publication in the National Bibliography;
detailed bibliographic data are available on the Internet at http://dnb.dnb.de .

Imprint:

Copyright © 2014 GRIN Verlag GmbH
Print and binding: Books on Demand GmbH, Norderstedt Germany
ISBN: 978-3-656-84962-9

***Jatropha curcas*: a potential genetic resource for herbal medicine and liquid bio-fuel**

Akhilesh Kumar

CSIR-National Botanical Research Institute, Rana Pratap Marg, Lucknow, 226 001, Uttar Pradesh, India

Abstract

Jatropha curcas L. (Euphorbiaceae) is a multipurpose perennial shrub/small tree, native to Mexico and Subtropical America, now grows naturally in most tropical areas of the world. It is an underutilized plant of multiple values. It is cultivated for seeds, having liquid biofuel potential and as live fence for the protection of agricultural crops. Various parts of the *J. curcas* are globally used for healthcare management of plants, human being and domesticated animals. Besides ethnomedicinal usages, this species have much other ethnobotanical, economic and ecological importance. Present paper deals with origin and distribution, taxonomic description, propagation and cultivation, utilization, pharmacological activities, phytochemical properties and future prospective of this species.

Keywords: Ethnobotany; Ethnomedicine; Euphorbiaceae; *Jatropha curcas*; Underutilized crop

1. Introduction

The word 'Jatropha' is derived from Greek words 'Jatros' and 'trope' (food/nutrition) which implies medicinal uses. The genus *Jatropha* belongs to family Euphorbiaceae and subfamily Acalyphoideae, and includes about 175 species. *Jatropha curcas* L. syn *Curcas purgans* Medik., *Ricinus americanus* Miller, *Castiglionia lobata* Ruiz & Pavon, *Jatropha edulis* Cerv., *J. acerifolia* Salisb., *Ricinus jarak* Thunb., *Curcas adansoni* Endl., *Curcas indica* A. Rich., and *Curcas curcas* (L.) Britton & Millsp etc. is the most important species of the genus. Linnaeus classified the plant in 1753 and gave it the botanical name *Jatropha curcas* (Heller 1996; Krishnan and Paramathma 2009; USDA plant database). It has 2n = 22 chromosomes (Soontornchainaksaeng and Jenjittikul 2003; Jha et al. 2007; Carvalho et al, 2008). It is commonly known as physic nut, purging nut, barbados nut, and nutmeg plant in English. Other vernacular names of *J. curcas* are pinhão manso, mundubi-assu (Brazil), pourghère (French), purgeernoot (Dutch), purgiernuss (German), purgeira, pinha˜o-manso

(Portuguese), fagiola d'India (Italian), galamaluca (Mozambique), kadam (Nepal), yu-lu-tzu (Chinese), habel meluk (Arab), kananaeranda, parvataranda (Sanskrit), safed arand, bagbherenda, jangaliarandi, ratanjot (Hindi), mogalierenda, erandagachh, ranayerendi, jamalgota, nepalamu, peddanepalamu, kadalamannku, kattamankku, adaluharalu, karnocchi, kattavanaka, jahazigzba, bongalibhotorna, borbandong (various parts of India), sabudam (Thai), bagani (Ivory Coast), butuje funfun (Nigeria), makaen (Tanzania), piñoncillo (Mexico), tempate (Costa Rica) and piñon (Guatemala) (Anonymous 1959; Heller 1996; Carvalho et al., 2008; Brittaine and Lutaladio 2010, Erinoso and Aworinde, 2012).

1.1 Origin and distribution

The origin of *J curcas* remains controversial as it can be found over a wide range of countries in Central and South America. It is native to Central America, but now grows naturally in most tropical areas of the world (Burkill 1994; Heller 1996; Openshaw, 2000, Fairless 2007). Portuguese introduced *J. curcas* in Asia and Africa as an oil yielding plant. In India it occurs in wild, semi wild and cultivated state in almost all biogeographical zones from the coastal areas to the outer Himalayan ranges (Anonymous, 1959).

1.2 Morphological description (See also Anonymous 1959; Singh, 1970; Heller 1996; Raju and Ezradanam, 2002; Bhattacharya et al. 2005; Achten et al 2008; Brittaine and Lutaladio 2010)

J curcas is a multipurpose perennial shrub/small tree of 3-6 m height. It may be evergreen or deciduous, depending on climate. It has a short tap root, robust laterals, and many fine tertiary roots. The stem is woody, erect, cylindrical, solid and branched. Branches are stout, green, and semi woody. Leaves are palmate and have 5 to 7 shallow lobes and are arranged in alternate with spiral phyllotaxis. Length and widths of leaves varies from 16 to 21 and 14 to 18 cm and are cauline and ramel, ex-stipulate, petiolate. Petioles are 12–19 cm long. Venation is multicostate, reticulate, and divergent type.

J curcas is monoecious and the terminal inflorescences contain unisexual greenish yellow 17-105 male and 2-19 female flowers in loose panicle of cymes. The ratio of male to female flowers ranges from 13:1 to 29:1. The inflorescence is composed by a main florescence and a distinct coflorescence. There are nodes on the upper pedicels of male (staminate) flowers, and no node on the upper pedicels of female (pistillate) flowers. The flowers are tiny (about 7 mm), unisexual, regular, petals are oblong and light green in colour, and sepals are quinquepartite. Androecium is absent in female flower, present in male flower with ten stamens. Stigma are six furcated, dorsifixed and introrse. Gynoecium is absent in male flowers, but present in female flowers and is tricarpellary, syncarpous with trilocular, and superior ovary. Flowers are pollinated by moths and bees. Fruits trilocullar, ellipsoidal, sudrupaceous. The exocarp remains fleshy until the seeds become mature, finally separating into three cocci. The fruit is 2.5-3.5 cm long to 2-2.5 cm wide. Seeds are black, oblong, 2.5 to 3 cm long and 1 cm thick, more or less spherical or ellipsoidal (Figure 1). Seed weight (10 seed) ranges from 53-77 g which contains 13.06-42.41 % oil content.

2. Propagation and cultivation

J. curcas can be easily propagated by seed, stem cutting, grafting, air layering, and tissue culture. Plantation of direct seedlings or pre-cultivated seedlings and direct planting of cuttings methods are suitable for large scale cultivation. In agro-forestry and intercropping systems direct seeding should be preferred over pre-cultivated plants, as the taproot of directly seeded plants is believed to penetrate in deeper soil layers. Cuttings of 25-120 cm from one year old branches can be planted directly in field or in nursery bed/polythene bags for first root development during rainy season. In case of generative propagation seeds can be pre-treated with cow-dung slurry (12 hours) or cold water (24 hours) before showing in nursery beds/ polythene bags filled with sandy-loam soil and compost (1:1 ratio). Seeds should be shown three months before rainy season in nursery and during rainy season in case of direct showing in field (Heller, 1996, Openshaw 2000, Kausik and Kumar 2006; Achten et al. 2008; Brittaine and Lutaladio 2010).

There is growing surge in cultivation of *J. curcas* worldwide, but it is still a wild plant and its basic agronomic properties and environmental effects have yet to be analysed. The limited knowledge of agronomic practices, diseases and insect pest management are major constrains in successful cultivation of *J. curcas* as a biofuel crop. Sandy and gravelly well-drained and aerated soils is most suitable for cultivation of *J curcas;* however, it can grow on large range of soil but unable to tolerate water logging conditions in heavy clay soils. Optimum soil pH for *J curcas* is 6 to 8; however, it can grow in soils from 5.5 to 9.0 pH range. It can grow in 250 to 3000 mm per annum rainfall regime, but at least 600 mm annual rainfall is required for flowering and fruiting. It can tolerate high temperature but cannot tolerate frost. 19.3^0 to 27.2^0 C is optimum temperature for *J curcas*. It can grow in marginal soils with low nutrient content, but in order to get commercial seed production appropriate fertilizer input is required. Organic fertilizer is better on marginal and degraded sites. Unfortunately, there is insufficient data on response to fertilizer application on seed production so it is not possible to make specific recommendations for fertilizer input under different agro-climatic conditions; more research is required in this regard. Field preparation for *J curcas* cultivation includes clearing of land, labelling and preparation of pits. Pits can be dug either manually (30-45 x 30-45 x 30-45 cm^3) or with drill machine (30-45 cm diameter and 1 m depth) before rainy season and refilled with soil, sand and fertilizer. *J curcas* may be planted at 2 x 2 or 3 x 3 m^2 distance in the field during rainy season and requires 1100-2500 saplings or seeds per hectare. Plants should be irrigated just after plantation and requires regular care and irrigation during first year. Weeding and pruning are important management practices for commercial seed production as former reduces competition and latter increases number of branches resulting more flowering and fruiting. Though, *J curcas* is not palatable to livestock, but there is need to protect the plantations in early stages of their development to prevent massive destruction by roaming grazers (Foidl et al. 1996; Heller 1996; Tiwari et al. 2007; Achten et al. 2008; Maes et al. 2009; Behera et al. 2010; Brittaine and Lutaladio 2010; Sop et al. 2011). Recently Paclobutrazol and Benzyladenine treatment has been reported to increase seed yield significantly (Ghose et al. 2011; Pan et al. 2011). Two kg Farm Yard Manure (organic manure)/ planting pit and nitrogen, phosphorous and potassium at 10 g, 20 g, and 10 g, respectively and Irrigation at 30 days interval is recommended in degraded soil (Singh et al. 2013). Harvesting and separation of seeds from fruits are done manually. Seeds should be dried to maintain moisture content up to 6-9% before oil extraction. The extraction of the oil can be done by various mechanical or chemical techniques (Achten et al. 2008).

As for as diseases and insect pests are concerned, *J curcas* is attacked by many pathogens and insects pests (Heller 1996; Grimm 1999; Shankar and Dhyani 2006; Ambika et al. 2007; Tewari et al. 2007; Latha et al. 2009; Pereira et al. 2009; Kumar et al 2010; Rao et al. 2011; Wu et al. 2011). Fruit feeding by *Scutellera perplexa* may led to premature fruit and seed abortion and reduction in yield (Sahai et al. 2011). Termites may cause severe damage at certain sites (Sop et al. 2011). Regular irrigation and fertilizer application in commercial monocultures is expected to enhance infestations of these insect pests and disease as observe in most of food and fibre crops, so the use of pesticides from the beginning of the plantings is recommended (Heller 1996; Achten et al. 2008; Fitt 2011; Sop et al. 2011).

3. Utilization

J curcas is emerging as an interesting multipurpose species within academic, civil society and policy makers. The seed oil can be easily converted into liquid biofuel which meets the American and European standards (Azam et al. 2005; Fairless 2007; Tiwari et al. 2007). Various parts of *J. curcas* can be used for healthcare management of plants, animals, and human being (Table 1). Besides biofuels and healthcare management, *J. curcas* is also useful to control soil erosion and improved water infiltration, to reclaim wasteland, phytoremediation of various contaminated soils, livestock barrier and land demarcation or live fence around agricultural fields, fuel wood, and support for vanilla, green manure, soil carbon sequestration and sustainable environmental development. Other economic products obtained from various parts of *J. curcas* are glycerol, soap, cosmetics, varnish, dye, molluscicide, pesticide, fertilizer, synthesis of silver nano-particles (Heller 1996; Mangkoedihardjo and Surahmida, 2008; Bar et al. 2009; Jameel et al. 2009; Sharma et al. 2009; Agamuthu et.al., 2010; Brittaine and Lutaladio 2010; Pandey et al. 2012; Warra 2012). Seeds are toxic for human (Kulkarni et al. 2005) and animal and are used as ordeal poisons for internal use in Africa (De Smet 1998). Seed cake may also be used for human consumption after treatment (Schuh and Schuh 2012).

3.1 Pharmacological activities

3.1.1. Anti-bacterial Activity

Acetone, chloroform, ethanol and methanol extracts of *J. curcas* root bark has been reported to inhibit the growth of both gram-positive (*Staphylococcus aureus*) and gram negative bacteria like *Klebsiella pneumonia, Pseudomonas aeruginosa, Salmonella typhimurium* and *Escherichia coli* (Naovi et al. 1991; Muanza et al. 1994; Tona et al. 1999; Sundari et al. 2011).

3.1.2. Anti-fungal activity

Various plant parts of *J. curcas* have antifungal activity against *Aspergillus fumigatus, A. flavus, A. niger, Bacillus subtillis, Phymatotrichopsis omnivora, Candida albicans* etc. which are responsible for many diseases in human being and plants (Naovi et al. 1991; Muanza et al. 1994; Kubmarawa et al. 2007; Hu et al 2011; Sundari et al. 2011).

3.1.3 Antiviral activity

The water extract of the branches of J curcas strongly inhibit the HIV-induced cytopathic effects with low cytotoxicity (Matsuse et al. 1998). Latex of J curcas possesses Inhibitory property against water melon mosaic virus (Tewari and Shukla 1982)

3.1.4 Anti-inflammatory activity

Topical application of *J curcas* root powder in paste form in mice and rats has been reported to possess anti-inflammatory activity by Mujumdar and Misar (2004).

3.1.5 Anti-oxidant activity

Root bark extract of *J. curcas* were capable of scavenging hydroxyl in a concentration dependent manner and have a stronger hydroxyl scavenging activity of compared with ascorbic acid (Sundari et al. 2011).

3.1.6. Coagulant and anticoagulant activities

Latex is a blood coagulant whereas dilute latex is anticoagulant. Curcin from seed produces deleterious effects to the blood (Osoniyi and Onajobi 2003).

3.1.7. Anti-diarrhoeal activity

The petroleum ether and methanol extract of *J curcas* roots shows anti-diarrhoeal activity in various species of albino mice (Mujumdar et al. 2000).

3.1.8. Pregnancy terminating effect

Pregnancy terminating effect of methanol, petroleum ether and dichloromethane extracts extract of *J curcas* fruits in rats have been scientific reported by Goonasekera et al. (1995).

3.1.9. Wound healing activity

Wound healing activities of stem bark of *J curcas* has been reported in literature (Villegas et al. 1997; Shetty et al. 2006; Igbinosa et al. 2009. Sachdeva et al. (2011) had scientifically evaluated wound healing potential of white soft paraffin base ointment containing 5% (w/w) and 10% (w/w) extract of stem bark of *J curcas* using incision and excision wound model in albino rats.

3.1.10. Insecticidal, larvicidal and anthelmintic activity

Various plant parts of *J curcas* have been reported to possess insecticidal and larvicidal and anthelmintic activity. Insect pests of stored grains (*Sitophilus zeamais* and *Rhyzorpertha dominica*) are susceptible to seeds and pericarps of *J curcas* (Silva et al. 2012). Ethanol extract of leaves of *J curcas* may be as useful for developing a safe and ecofriendly therapeutic agent to combat the problems of tick *Rhipicephalus(Boophilus) annulatus* and tick-borne diseases (Juliet et al. 2012). *J curcas* is a potential source of herbal mosquito control agent. Larvicidal activities methanol extract of leaves, crude protein extract and purified toxin, Jc-SCRIP, from the seed coat of *J curcas* has the larvicidal potential against *Aedes aegypti, Anopheles arabiensis, Aedes aegypti* and *Culex quinquefasciatus* (Karmegam et al. 1996; Rahuman et al. 2008; Sakthivadivel and Daniel 2008; Aina et al. 2009; Cantrell et al. 2011; Zewdneh et al 2011; Tomass et al. 2011; Kalimuthu et al. 2011; Nuchsuk et al. 2012). Aqueous extract of leaves of *J curcas* possesses anthelmintic activity against *Pheritima poshtuma* (Ahirrao et al. 2009, 2011).

4. Phytochemicals

The leaf, bark and latex of *Jatropha* contains alkaloids such as jatrophine, jatropham, curcacycline A, curcain, tannins, glycosides, flavonoids and sapogenins with anti-cancerous properties (van den Berg et al 1995, Thomas et al., 2008; Debnath and Bisen, 2008). The seeds of *J.curcas* contains some toxic compounds such a protein (curcin) and phorbol-esters diterpenoids (King et al. 2009). The diterpenes isolated from *Jatropha* species belongs to rhamnofolane, daphnane, lathyrane, tigliane, dinorditerpene, deoxy preussomerin and pimarane skeletal structures and the majority of the diterpenes exhibited cytotoxic, antitumor and antimicrobial activities in vitro. Jatrophone, spruceanol and jatrophatrione exhibited antitumor properties against P338 lymphocytic leukaemia and japodagrol against KB carcinoma cells. Curcusone exhibited anti-invasive effects against cholangiocarcinoma cells. The phorbol esters (Jatropha factor C1–C6) and jatropherol exhibited insect deterrent/cytotoxic properties. Jatrophalactam, faveline derivatives, multifolone, curcusone, jatrophone derivatives etc. have shown in-vitro cytotoxic activity. Japodagrin, jatrogrossidione derivatives and jatropholone derivatives exhibited antimicrobial activities.

Jatropha diterpenoids having a wide spectrum of bioactivity could form lead compounds or could be used as templates for the synthesis of new compounds with better biological activity for utilization in the pharmaceutical industries (Devappa et al. 2011). Three deoxypreussomerins, palmarumycins CP1, JC1 and JC2, have been isolated from the stems of *J curcas* (Ravindranatha et al. 2004) which possess a wide range of biological properties including antibacterial, antifungal, herbicidal, antibiotic and antitumor activities (Wipf et al. 2001). *J curcas* seed oil chemically consists of triacylglycerol with linear fatty acid chain. Palmitic acid, stearic acid, oleic acid and linolic acid, lauric acid, miristic acid, arachidic acid, arachidolic acid and behenic acid are some important fatty acids present in J curcas seed oil (Adebowale and Adedire 2006; de Oliveira et al. 2009). Ling et al. (2010) identified a new chemical compound jatrophasin A (3,4,4',5'-tetrahydroxyl-3'-methoxyl-bisepoxylignan) with strong anti-oxidative activity from the seeds of *J curcas*.

5. Future perspective

Large scale seed production of *J curcas* remains the single most important issue that will ultimately decide the success of this crop at commercial level. High seed productivity and oil content are desired for commercial utilization of this crop for biofuel production. The challenges of developing viable market for *J curcas* are -

* Besides biofuel, research on medicinal, plant protecting and other economic potential of various plant parts of *J curcas* for development of new pharmaceuticals and plant protectants of herbal origin.
* Economic analysis of the biogas production potential of husk and seed cake.
* Economic analysis of fertilizer value of the seed cake.
* Identification of sufficient quantity and quality of available land for large scale commercial cultivation of *J curcas* which do not compete with food production.
* Identification of suitable accessions of *J curcas* for various types of degraded lands and agro climatic zones.
* Identification of suitable practice of packages (irrigation, nutrient management requirement and optimum use of organic, chemical and/or biofertelizer etc.) for cultivation of *J curcas* in various types of degraded lands and agro climatic zones.
* Practice of apiculture in *J curcas* field to promote pollination resulting good seed yield.
* Though influence of pruning on growth and dry mass partitioning has been analysed by Rajaona et al. (2011), but its effect on seed yields has yet to be analysed.

* Intercropping of *J curcas* with non edible crops like essential oils to increase income per unit land area.

* Building agencies to facilitate trade of seed between smallholder farmers and industries using the seed for oil extraction and processing for production of value added products including biodiesel.

* Increasing investment in *Jatropha* research projects for development of pharmaceutical and plant protectants (insecticide and pesticide for crops and stored grains) along with projects ensuring commercial seed production on various types of waste land and efficient seed oil extraction methods.

* Economic analysis of use of seed oil for agricultural instruments (tube well, tractors etc.), lighting and soap production in rural areas.

* Socioeconomic studies on how *J curcas* can aid development in rural areas.

Acknowledgements

Authors are thankful to the director, CSIR-National Botanical Research Institute for providing necessary facilities.

References

Achten WMJ, Verchot L, Franken YJ, Mathijs E, Singh VP, Aerts R, Muys B (2008) *Jatropha* bio-diesel production and use. Biomass Bioenergy 32:1063–1084

Adebowale KO, Adedire CO (2006) Chemical composition and insecticidal properties of the underutilized *Jatropha curcas* seed oil. African J Biotechnol 5:901-906

Agamuthu P, Aioye OP, Aziz AA (2010) Phytoremediation of soil contaminated with used lubricating oil using *Jatropha curcas*. J Hazard Matter 179: 891–894

Ahirrao RA, Patel MR, Sayyed H, Patil JK, Suryawanshi HP, Tadavi SA (2011) In vitro anthelmintic property of various herbal plants extracts Against *Pheritima posthuma*. Pharmacologyonline 2: 542-547.

Ahirrao RA, Pawar SP, Borse LB, Borse SL, Desai SG, Muthu AK (2009). Anthelmintic activity of leaves of *Jatropha curcas* Linn and *Vitex negundo* Linn. Pharmacologyonline 1:276-279.

Aina SA, Banjo AD, Lawal OA, Jonathan K (2009) Efficacy of some plant extracts on *Anopheles gambiae* mosquito larvae. Acad J Entomol 2:31–35

Ambasta SP (1994) Useful plants of India. CSIR, New Delhi

Ambika S, Manoharan T, Stanley J, Preetha G (2007) Scutellarid pests of *Jatropha* and their management. Ann Pl Protec Sci 2007:370-375

Anonymous (1959) The wealth of India- Raw material. Volume 4, CSIR, New Delhi 239-397.

Azam MM, Waris A, Nahar NM (2005) Prospects and potential of fatty acid methyl esters of some non-traditional seed oils for use as biodiesel in India. Biomass Bioenergy 29:293–302.

Bar H, Bhui DK, Sahoo GP, Sarkar P, De SP, Misra A (2009) Green synthesis of silver nano particles using latex of *Jatropha curcas*. Colloids Surf A Physicochem Eng Asp 339:134–139

Baral SR, Kurmi PP (2006) A compendium of medicinal plants in Nepal. Mrs Rachana Sharma publication, Kathmandu

Behera SK, Srivastava P, Tripathi R, Singh JP, Singh N (2010) Evaluation of plant performance of Jatropha curcas L. under different agro-practices for optimizing biomass – A case study. Biomass Bioenergy 34: 30-41

Bhattacharya A, Datta K, Datta SK (2005) Floral biology floral resource constraints and pollination limitations in Jatropha curcas L. Pakistan J Biol Sci 8: 456–460.

Brittaine R, Lutaladio NB (2010) Jatropha: a smallholder bioenergy crop: the potential for pro-poor development. Integrated crop management Vol. 8. Food and Agriculture Organization of the United Nations, Rome

Burkill HM (1994) The useful plants of West Tropical Africa (Families E-J) Royal Botanical Gardens Kew 90-94

Cantrell C, Ali A, Duke SO, Khan I (2011) Identification of mosquito biting deterrent constituents from the Indian folk remedy plant *Jatropha curcas*. J Med Entomol 48:836–845

Cartaxo LS, de Almeida Souza MM, de Albuquerque UP (2010) Medicinal plants with bioprospecting potential used in semi-arid northeastern Brazil. J Ethnopharmacol 131:326–342

Carvalho CR, Clarindo WR, Praca MM, Araujo FS, Carels N (2008) Genome size, base composition and karyotype of *Jatropha curcas* L. an important biofuel plant. Plant Science 174: 613–617

Chowdhury MSH, Koike M (2010) Towards exploration of plant-based ethno-medicinal knowledge of rural community: basis for biodiversity conservation in Bangladesh. New Forests, 40:243–260

de Oliveiraa J, Leitea PM, de Souzaa LB, Melloa VM, Silvab EC, Rubima JC, Meneghettib SMP, Suareza PAZ (2009) Characteristics and composition of *Jatropha gossypiifolia* and

Jatropha curcas L. oils and application for biodiesel production. Biomass Bioenergy 33:449– 453

De Smet PAGM (1998) Traditional pharmacology and medicine in Africa Ethnopharmacological themes in sub-Saharan art objects and utensils. J Ethnopharmacol 63:1–179

Debnath M, Bisen PS (2008) Jatropha curcas L., A Multipurpose Stress Resistant Plant with a Potential for Ethnomedicine and Renewable Energy. Current Pharmaceutical Biotechnology 9: 288-306

Deshmukh RR, Rathod VN, Pardeshi VN (2011) Ethnoveterinary medicines from Jalna distict of Maharastra state. Indian J Tradit Know 10:344-348

Devappa RK, Makkar HPS, Becker K (2011) Jatropha Diterpenes: a Review. J Am Oil Chem Soc 88:301–322

Erinoso SM, Aworinde DO (2012) Ethnobotanical survey of some medicinal plants used in traditional health care in Abeokuta areas of Ogun State, Nigeria. African J Pharmacy Pharmacol 6(18):1352-1362

Fairless D (2007) Biofuels: The little shrub that could maybe. Nature 449:652-655

Fitt GP (2011) Critical issues in pest management for a future with sustainable biofuel cropping. Curr Opinion Environ Sustain 3:71–74

Foidl N, Foidl G, Sanchez M, Mittelbech M, Hackel S (1996) *Jatropha curcas* L as a source for the production of biofuels in Nicaragua. Bioresour Technology 58:77– 82

Ghos A, Chikara J, Chaudhary DR (2011) Diminution of economic yield as affected by pruning and chemical manipulation of *Jatropha curcas* L. Biomass Bioenergy 35: 1021-1029.

Goonasekera MM, Gunawardana, VK, Jayasena K, Mohammed S.G, Balasubramaniam S (1995) Pregnancy terminating effect of *Jatropha curcas* in rats. Journal of Ethnopharmacology 47:117-123

Gradé JT, Tabuti JRS, Damme PV (2009) Ethnoveterinary knowledge in pastoral Karamoja, Uganda. J Ethnopharmacol 122: 273–293

Grimm C (1999) Evaluation of damage to physic nut (*Jatropha curcas*) by true bug. Entomolagia Expermentalish et Applica 92: 127-136

Heller J (1996) Physic nut – *Jatropha curcas* L. promoting the conservation and use of underutilized and neglected crop. Ph D dissertation. Institute of plant genetic research, Gatersleben, germany; International Plant Genetic Resource, Institute, Rome, Italy.

Hu P, Wang AS, Engledow AS, Hollister EB, Rothlisberger KL, Matocha JE, Zuberer DA, Provin TL, Hons FM, Gentry TJ (2011) Inhibition of the germination and growth of *Phymatotrichopsis omnivora* (cotton root rot) by oilseed meals and isothiocyanates. Applied Soil Ecol 49:68-75

Idu M, Timothy O, Onyibe HI, Comor AO (2009) Comparative Morphological and Anatomical Studies on the Leaf and Stem of some Medicinal Plants: *Jatropha curcas* L. and *Jatropha tanjorensis* J.L. Ellis and Saroja (Euphorbiaceae). Ethnobotanical Leaflets 13: 1232-1239

Igbinosa OO, Igbinosa EO, Aiyegoro OA (2009) Antimicrobial activity and phytochemical screening of stem bark extracts from *Jatropha curcas* (Linn). Afr J Pharm Pharacol 3(2): 58-62

Jameel S, Abhilash PC, Singh N, Sharma PN (2009) *Jatropha curcas*: a potential crop for phytoremediation of coal fly ash. J Hazard Matter 172:269-275

Jha TB, Mukherjee P, Datta MM (2007) Somatic embryogenesis in *Jatropha curcas* Linn., an important biofuel plant. Plant Biotechnol 1:135–140

Juliet S, Ravindran R, Ramankutty SA, Gopalan AKK, Nair SN, Kavillimakkil AK, Bandyopadhyay A, Rawat AKS, Ghoshe S (2012) *Jatropha curcas* (Linn) leaf extract –a possible alternative for population control of *Rhipicephalus(Boophilus) annulatus*. Asian Pacific J Trop Disease 2:225–229

Kalimuthu K, Kadarkaral M, Savariar V, Siva K (2011) Larvicidal efficacy of *Jatropha curcas* and bacterial insecticide, Bacillus thuringiensis, against lymphatic filarial vector, Culex quinquefasciatus Say (Dipteria: Culicidae). Parasitol Res 109 (5):1251-1257

Karmegam J, Sakthivadivel M, Daniel T (1996) Indigenous plant extracts as larvicidal agents against *Culex quinquefasciatus* say. Bioresour Technol 59:137–140

Kaushik N, Kumar S (2006) *Jatropha curcas* L. silviculture and uses. Agrobios publication Jodhpur

King AJ, He W, Cuevas JA, Freudenberger, Ramiaramanana D and Graham IA.2009. Potential of Jatropha curcas as a source of renewable oil and animal feed. J Experimental Bot 60: 2897-2905

Krishnan PR, Paramathma M (2009) Potentials and *Jatropha* species wealth of India. Current Science 97 (7): 1000-1004

Kubmarawa D, Ajoku GA, Enwerem NM, Okorie DA (2007) Preliminary phytochemical and antimicrobial screening of 50 medicinal plants from Nigeria. African J Biotech 6:690-1696

Kulkarni ML, Sreekar H, Keshavamurthy KS, Shenoy N (2005) *Jatropha curcas* - poisoning. Indian J Pediatrics 72:75-76

Kumar A, Prakash P, Singh N (2010) Maconellicoccus hirsutus (Green) Infestation on Jatropha curcas L saplings and its possible management through herbal pesticides. "Fourth International Conference on "Plant and Environmental Pollution" (ICPEP-4) held at NBRI, Lucknow on Dec. 08-11, 2010

Kunwar RM, Uprety Y, Burlakoti C, Chowdhary CL, Bussmann RW (2009) Indigenous Use and Ethnopharmacology of Medicinal Plants in Far-west Nepal. Ethnobot Res Appli 7:005-028

Lans C (2007) Comparison of plants used for skin and stomach problems in Trinidad and Tobago with Asian ethnomedicine. J Ethnobiol Ethnomed 3:3

Lans C, Harper T, Georges K, Bridgewater E (2001) Medicinal and ethnoveterinary remedies of hunters in Trinidad. BMC Complement Alter Med 1:10

Latha P, Prakasham V, Kamlakannan A (2009) First report of *Lasiodiplodia theobromea* (Pat.) Griffon & Maubl causing root rot and collar rot disease in physic nut (*Jatropha curcas* L) in India. Astralasian Plant Disease notes 4:19-20

Ling LI, Xin-luan WANG, Xiao-fan LI, Nai-li WANG (2010) A New Compound with Anti-oxidative Activity from Seeds of *Jatropha curcas*. Chinese Herbal Medicines, 2(4): 245-247.

Longuefosse, JL, Nossin E (1996) Medical ethnobotany survey in Martinique. J Ethnopharmacol 53:117-142

Luseba, D, Van der Merwe D (2006) Ethnoveterinary medicine practices among Tsonga speaking people of South Africa. Onderstepoort J Vet Res 73:115–122

Maes WH, Trabucco A, Achten WMJ, Muys V (2009) Climatic growing conditions of *Jatropha curcas* L. Biomass Bioenergy 33: 1481– 1485

Manandhar NP (2002) Plant and People of Nepal. Timber Press Inc. Portland, Oregon

Mangkoedihardjo S, Surahmida (2008) *Jatropha curcas* L for phytoremediation of lead and cadmium polluted soil. World Appl Sci J 4:519-522

Matsuse IT, Lim YA, Hattori M, Correa M, Gupta MP (1998) A search for anti-viral properties in Panamanian medicinal plants: The effects on HIV and its essential enzymes. J Ethnopharmacol 64:15-22

McGaw LJ, Eloff JN (2008) Ethnoveterinary use of southern African plants and scientific evaluation of their medicinal properties. J Ethnopharmacol 119:559–574

Mesfin A, Giday M, Animut A, eklehaymanot T (2011) Ethnobotanical study of antimalarial plants in Shinile District, Somali Region, Ethiopia, and in vivo evaluation of selected ones against *Plasmodium berghei*. J. Ethnopharmacol 139 :221-227 doi:10.1016/j.jep.2011.11.006

Moshi MJ, Otieno DF, Weisheit A (2012) Ethnomedicine of the Kagera Region, north western Tanzania. Part 3: plants used in traditional medicine in Kikuku village, Muleba District. J Ethnobiol Ethnomed 8:14

Muanza DN, Kim BW, Euler KL, Williams L (1994) Antibacterial and antifungal activities of nine medicinal plants from Zaire. Int J Pharmacogn 32:337–345

Mujumdar AM, Misar AV (2004) Anti-inflammatory activity of *Jatropha curcas* roots in mice and rats. J Ethnopharmacol 90:11-15

Mujumdar AM, Upadhye AS, Misar AV (2000) Studies on antidiarrhoeal activity of *Jatropha curcas* root extract in albino mice. J Ethnopharmacol 70:183-187

Namsa ND, Tag H, Mandal M, Kalita P, Das AK (2009) An ethnobotanical study of traditional anti-inflammatory plants used by the Lohit community of Arunachal Pradesh, India. J Ethnopharmacol 125 (2): 234-245

Naovi SAH, Khan MSY, Vohora SB (1991) Anti-bacterial, anti-fungal and anthelmintic investigations on Indian medicinal plants. Fitoterapia 62:221-228

Nath KK, Deka P, Borthakur SK (2011) Traditional remedies of joint pain in Assam. Indian J Tradit Know 10:568-571

Nuchsuk C, Wetprasit N, Roytrakul S, Ratanapo S (2012) Larvicidal activity of a toxin from the seeds of *Jatropha curcas* Linn. against *Aedes aegypti* Linn. and *Culex quinquefasciatus* Say. Trop Biomed 29(2):286–296

Offiah, N V, Sunday M, Elisha, I L, Makoshi, M S, Gotep, JG, Dawurung CJ., Oladipo, OO, Lohlum, AS, Shamaki D (2011) Ethnobotanical survey of medicinal plants used in the treatment of animal diarrhoea in Plateau State, Nigeria. BMC Vet Res 7:36

Openshaw K (2000) A review of *Jatropha curcas*: an oil plant of unfulfilled promise. Biomass Bioenergy 19:1-15

Osoniyi O, Onajobi F (2003) Coagulant and anticoagulant activities in *Jatropha curcas* latex. J Ethnopharmacol 89:101-105

Pan BZ, Xu ZF (2011) Benzyladenine treatment significantly increases the seed yield of the biofuel plant *Jatropha curcas*. J Plant Growth Regul 30:166–174

Pandey VC, Singh K, Singh JS, Kumar A, Singh B, Singh RP (2012) *Jatropha curcas*: A potential biofuel plant for sustainable environmental development. Rene Sus Energy Reviews 16:2870–2883

Pereira OL, Dutra DC, Dias LAS (2009) *Lasiodiplodia theobromea* is the causal agent of a damaging root and collar rot disease on the biofuel plant *Jatropha curcas* L in Brazil. Astralasian Plant Disease notes 4:120-123

Pragada PM, Rao GMN (2012) Ethnoveterinary medicinal practices in tribal regions of Andhra Pradesh, India. Bangladesh J Plant Taxon 19:7-16

Rahuman AA, Gopalakrishnan G, Venkatesan P, Geetha K (2008) Larvicidal activity of some Euphorbiaceae plant extracts against *Aedes aegypti* and *Culex quinquefasciatus* (Diptera: Culicidae). Parasitol Res 102:867–873

Rajaona AM, Brueck H, Asch F (2011) Effect of pruning history on growth and dry mass partitioning of jatropha on a plantation site in Madagascar. Biomass Bioenergy 35:4892-4900.

Raju AJS, Ezradanam V (2002) Pollination ecology and fruiting behaviour in a monoecious species *Jatropha curcas* L. (Euphorbiaceae). Current Science 83:1395-1398

Rao SC, Kumari MP, Wani SP, Marimuthu S (2011) Occurrence of black rot in *Jatropha curcas* L. plantations in India caused by *Botryosphaeria dothidea*. Current Science 100: 1547-1549

Ravindranath N, Reddy MR, Mahender G, Ramu R, Kumar KR, Das B (2004) Deoxypreussomerins from *Jatropha curcas*: are they also plant metabolites? Phytochemistry 65:2387–2390

Sachdeva K, Garg P, Singhal M, Srivastava B (2011) Wound healing potential of extract of *Jatropha curcas* L. (Stem bark) in rats. Phocg J 3 (25): 67-72

Sahai K, SrivastavaV, Rawat KK (2011) Impact assessment of fruit predation by *Scutellera perplexa* Westwood on the reproductive allocation of *Jatropha*. Biomass Bioenergy 35: 4684–4689.

Saikia AP, Ryakala VK, Sharma P, Goswami P, Bora U (2006) Ethnobotany of medicinal plants used by Assamese people for various skin ailments and cosmetics. J Ethnopharmacol 106:149–157

Sakthivadivel M, Daniel T (2008) Evaluation of certain insecticidal plants for the control of vector mosquito vis. *Culex quinquefaciatus, Anopheles staphensi* and *Aedes aegypti*. Appl Entomol Zool 43:57–63

Schuh AJ, Schuh PA (2012) Production of human food from *Jatropha* and other biological. United State Patent, Patent No. US 8,137, 717 B1

Shanker, Chitra and Dhyani, S. K., (2006) Insect pests of *Jatropha curcas* L. and the potential for their management. Current Science, 91:162-163

Sharma DK, Pandeya AK, Lata (2009) Use of *Jatropha curcas* hull biomass for bioactive compost production. Biomass Bioenergy 33:159–162

Shetty S, Udupa SL, Udupa AL, Vollala VR, 2006. Wound healing activities of bark extract of *Jatropha curcas* Linn in albino rats. Saudi Med J 27:1473-1476

Silva GN, Faroni LRA, Sousa AH, Freitas RS (2012) Bioactivity of *Jatropha curcas* L. to insect pests of stored products. J Stored Prod Res 48:111–113

Singh RP (1970) Structure and development of seeds in Euphorbiaceae, *Jatropha* species. Beitr Biol Pflanz 47:79-90

Singh B, Singh K,Rao GR, Chikara J, Kumar D, Mishra DK, Sakia SP, Pathre UV, Raghuvansh N, Rahi TS, Tuli R (2013) Agro-technology of *Jatropha curcas* for diverse environmental conditions in India. Biomass Bioenergy 48:191-202

Soontornchainaksaeng P, Jenjittikul T (2003) Karyology of Jatropha (Euphorbiaceae) in Thailand. Thai Forest Bull 31:105–112

Sop TK, Kagambega FW, Bellefontaine R, Schmiedel U, Thiombiano A (2011) Effects of organic amendment on early growth performance of *Jatropha curcas* L. on a severely degraded site in the Sub-Sahel of Burkina Faso. Agroforest Syst DOI 10.1007/s10457-011-9421-4

Srivastava A, Patel SP, Mishra RK, Vashistha RK, Singh A, Puskar A (2012) Ethnomedicinal importance of the plants of Amarkantak region, Madhya Pradesh, India. International J Med Arom Plants 2:53-59

Sundari J, Selvaraj R., Rajendra Prasad N (2011) Antimicrobial and antioxidant potential of root bark extracts from *Jatropha curcas* (Linn). J Pharmacy Res 4(10):3743-3746

Tabuti JRS, Dhillion SS, Lye KA (2003) Ethnoveterinary medicines for cattle (*Bos indicus*) in Bulamogi county, Uganda: plant species and mode of use. J Ethnopharmacol 88:279–286

Tangjang S, Namsab ND, Arana C, Litin A (2011) An ethnobotanical survey of medicinal plants in the Eastern Himalayan zone of Arunachal Pradesh, India. J Ethnopharmacol 134:8–25

Tewari JP, Dwivedi HD, Pathak M, Srivastava SK (2007) Incidence of a mosaic disease in *Jatropha curcas* L. from Eastern Uttar Pradesh. Current Science 93:1048-1049

Tewari JP, Shukla IK (1982) Inhibition of infectivity of two strains of water melon mosaic virus by latex of some angiosperm. Geobios **9**: 124-126

Thomas R, Sah NK, Sharma PB (2008) Therapeutic biology of *Jatropha curcas*: a mini review. Curr Pharma Biotechnol, 9: 315-324

Tiwari AK, Kumar A, Raheman H (2007) Biodiesel production from Jatropha (*Jatropha curcas*) with high free fatty acids: an optimized process. Biomass Bioenergy 31:569–75

Tomass Z, Hadis M, Taye A, Mekonnen Y, Petros B (2011) Larvicidal effects of Jatropha curcas L. against Anopheles arabiensis (Diptera: Culicidea).. Momona Ethiopian Journal of Science, 3, 52-64.

Tona L, Kambu K, Mesia K, Cimanga K, Aspers S, de Bruyne T, Pieters L, Totte J (1999) Biological screening of traditional preparations from some medicinal plants used as antidiarrhoeal in Kinshasa, Congo. Phytomedicine 6:59–66

Umapriya T, Rajendran A, Arvindhan V, Thomoa B, Maharajan M (2011) Ethnobotany of Irular tribe in Palamalai Hills, Coimbatore, Tamilnadu. Indian J Nat Prod Resour, 2 (2):250-255

Upadhyay B, Parveen, Dhaker, AK, Kumar A, (2010). Ethnomedicinal and ethnopharmaco-statistical studies of Eastern Rajasthan, India. J Ethnopharmacol 129: 64-86

van den Berg AJJ, Horsten SFAJ, Bosch JJK, Kroes BH, Beukelman CJ, Leeflang BR, Labadie RP (1995) Curcacycline A- a novel cyclic octapeptide isolated from the latex of *Jatropha curcas* L. FEBS Letters 358:215-218

Villegas LF, Fernández ID, Maldonado H, Torres R, Zavaleta A, Vaisberg AJ, Hammond GB (1997) Evaluation of the wound-healing activity of selected traditional medicinal plants from Perú. J Ethnopharmacol 55:193-200

Wagh VV, Jain AK, Kadel C (2011) Ethnomedicinal plants used for curing dysentery and diarrhoea by tribes of Jhabua district (Madhya Pradesh). Indian J Nat Prod Resour 2 (2):256-260

Warra AA (2012) Cosmetic potentials of physic nut (*Jatropha curcas* Linn.) seed oil: A review. Am J Sci Ind Res 3: 358-366

Watt JM, Breyer-Brandwijk, MG (1962) The medicinal and poisonous plants of Southern and Eastern Africa. 2nd Edition. Livirigstone Ltd., London

Wipf P, Jung JK, Rodriguez S, Lazo JS (2001) Synthesis and biological evaluation of deoxypreussomerin A and palmarumycin CP1 and related naphthoquinone spiroketals. Tetrahedron 57:283–296

Wu YK, Ou G, Yu J (2011) First report of *Nectria haematococca* causing root rot disease of physic nut (*Jatropha curcas*) in China. Australasian plant disease notes 6: 39-42

Zewdneh T, Mamuye H, Asegid T, Yalemtsehay M, Beyene P (2011) Larvicidal effects of *Jatropha curcas* L. against *Anopheles arabiensis* (Diptera: Culicidae). MEJS 3:52–64

Fig. 1 Morphology of *Jatropha curcas* L. A plant having leaves and fruits, B stem, C leaves, D flower, E immature fruits, F matuure fruits, G seeds, H crude seed oil

Table 1: Traditional medicinal uses of various parts of *Jatropha curcas* L

S. N.	Plant part used	Medicinal uses	References
1	Leaves	Ulcers, septic gums, cuts, wounds, burns, itched and blistered skin, ulcer, purgative, pneumonia, rubifacient, lactagogue, mastitis, koilonichia, insecticidal, piles, diarrhoea, urinary infection, intestinal parasites, Stroke, thrombosis, toothache, pains, healing, haemostatic, malaria, hypertention,	Anonymous 1954; Ambasta, 1986; Longuefosse and Nossin 1996; Lans et al. 2001; Saikia et al. 2006; Lans 2007; Namsa et al. 2009; Brittaine and Lutaladio 2010; Upadhyaya et al. 2010; Cartaxo et al 2010, Tangjang et al. 2011, Offiah et al. 2011; Moshi et al. 2012; Pragada and Rao 2012, Srivastava et al 2012
2	Stem and stem bark	Toothache, gum problems, inflammation, pyorrhoea, Infectious diseases, tympani, sexually transmitted diseases, urinary infection, diarrhoea, indigestion, rheumatism, leprosy, fever, jaundice, gonorrhoea, pneumonia	Namsa et al. 2009; Brittaine and Lutaladio 2010; Upadhyaya et al. 2010; Offiah et al. 2011; Deshmukh et al. 2011; Umapriya et al. 2011; Srivastava et al 2012; Pragada and Rao 2012,
3	Latex/ plant sap /extract	Stop bleeding, reduced the blood clotting time, snake-bites, infected sores, treating newborns umbilical cords, coughs, mouth and throat sores, stroke, thrombosis, toothache, stomach-ache, pains in general, healing, haemostatic, HIV-AIDS, tumour, wound healing, allergies, burns, cuts and wounds, inflammation, leprosy and other dermatological disorders.	Anonymous, 1959; Ambasta, 1986; Lans et al. 2001, Manandhar 2002; Baral and Kurmi 2006; Kunwar et al. 2009; Idu et al. 2009; Cartaxo et al 2010; Chowdhury and Koike 2010; Upadhyaya et al. 2010;, Nath et al. 2011, Umapriya et al. 2011
4	Root/ root bark	Inflammation, external parasite, gout, rheumatism, diarrhoea, dysentery, gonorrhoea, east coast fever '*makebe*'.	Ambasta, 1986; Tabuti et al. 2003; Baral and Kurmi 2006; Brittaine and Lutaladio 2010;, Deshmukh et al. 2011, Wagh et al. 2011, Nath et al. 2011,
5	Flower	Diarrhoea	Offiah et al. 2011,
6	Fruit	Stroke, toothache, control of reproduction, inflammation, numbness after bug sting, to clean mother's and baby's blood during the pregnancy	Watt and Breyer-Brandwijk, 1962, Gradé et al. 2009,
7	Seeds/ seed oil	Eczema, skin diseases, rheumatic pain, purgative, gout, arthritis and jaundice, wound-healing, fractures, burns, boils, malaria, constipation, gonorrhoea, syphilis	Anonymous 1959; Ambasta 1986; Luseba and Van der Merwe, 2006; McGaw and Eloff 2008; Kunwar et al. 2009; Brittaine and Lutaladio 2010; Mesfin et al 2011, Erinoso and Aworinde 2012,